MASTER OF THE FLOW

A Complete Guide to Modern Irrigation Techniques,

by LILLIAN SAVAGE

retained without approval

from the publisher or creator.

TABLE OF CONTENTS

(Chapter 10)
Conclusion
Key Terms and Definitions
Glossary

Introduction

Irrigation, the regulated application of water to plants and crops, is an ancient method that has been crucial in the development of agriculture and civilization. Irrigation, which is the foundation of agricultural output, has developed from simple methods to complex systems that guarantee appropriate water distribution and crop health. This thorough book explores the fundamental significance of irrigation and charts the development of irrigation methods, illuminating how ingenuity and necessity have advanced this area.

Understanding Irrigation's Importance

The "elixir of life," or water, is commonly referred to as the lifeblood of agriculture. Any agricultural enterprise depends on the availability and effective use of water to succeed. To overcome the drawbacks of erratic rainfall and provide a steady supply of water for crops, irrigation is used. Improved crop yields, more food production, and increased agricultural sustainability are all a result of this water supply consistency.

In addition to traditional farming, irrigation is essential to horticulture, landscaping, and the upkeep of green spaces. Effective irrigation techniques are especially important in areas with limited water resources. Irrigation decreases water waste and lowers the risk of soil erosion by providing water directly to the root zones of plants.

Additionally, irrigation helps farmers grow a variety of products and produce surpluses, which promotes rural development and reduces poverty.

This surplus can be sold in markets, creating a new revenue stream and fostering economic expansion. The development of rural infrastructure and services as a result of irrigation-driven agricultural activities frequently improves the quality of life for communities.

The Development of Irrigation Methods

Early Irrigation Methods: Wells to Channels

Ancient civilizations including the Egyptians, Sumerians, and Indus Valley peoples are where irrigation first appeared. These early farmers used straightforward but efficient methods to harness water for their crops. One of the first techniques was creating earthen channels to direct water from rivers and streams into farms. These channels allowed for controlled flooding by directing water to the crops.

The construction of wells became a standard practice as cultures developed. Wells tapped into underground water supplies, supplying a steady supply of water even in parched areas. Although wells required a lot of human labor and were manually operated, they represented a tremendous advance in irrigation technology.

Persian qanats and Roman aqueducts were examples of innovations from antiquity.

Irrigation innovation was still thriving in antiquity. It is said that the Persians invented the brilliant underground irrigation system known as qanats. Through a network of mildly sloping tunnels, Qanats drew water from underground sources. This made it possible to distribute water effectively and stopped water from evaporating.

The Romans, who were recognized for their skill in engineering, built complex aqueducts to transport water across great distances. These aqueducts played a crucial role in bringing water to urban areas and agricultural territories while demonstrating the art of gravity-based water conveyance.

Norias and water wheels are examples of medieval innovations.

The Middle Ages saw the invention of cutting-edge water-lifting machinery like norias and water wheels. Water was raised from rivers and distributed to fields using norias, big wooden water wheels with buckets. Irrigation systems and grinding mills were powered by water wheels, which were frequently turned by rivers in motion.

Sprinklers, drip irrigation, and Beyond Modern Irrigation Systems

Irrigation practices underwent a paradigm shift throughout the Industrial Revolution. Modern materials and mechanization increased the efficiency and adaptability of irrigation systems. By supplying water through pressurized pipes and spraying it over crops in predetermined patterns, sprinkler systems revolutionized irrigation. This reduced water waste and made the application more accurate.

Another development is drip irrigation, which uses a system of tubes and emitters to supply water straight to the base of plants. This extremely effective technique saves water and prevents surface runoff, making it especially ideal for greenhouse farming and areas with limited water supplies.

Conclusion

Irrigation techniques have developed from their simple origins as channels and wells into intricate systems that guarantee the sustainability of agriculture and landscapes. It is impossible to stress how crucial irrigation is to environmental preservation, economic development, and food production. It becomes clear as we follow the development of irrigation that ingenuity and need have continually advanced the industry. We will delve into the complexity of drip irrigation, sprinkler systems, and smart irrigation technologies, all of which contribute to a greener and more sustainable future, as we examine these contemporary approaches in greater detail in the next chapters.

Foundations of Irrigation (Chapter 1)

A variety of methods, systems, and components are used in the multidimensional field of irrigation to effectively distribute water to plants and crops. We will examine the key concepts that create the framework for comprehending irrigation in this chapter. We'll explore the numerous irrigation system types, the fundamental parts that make up an irrigation system, and the various variables that affect judgments about irrigation strategies.

Different Irrigation System Types

Different irrigation system types have been developed to accommodate various terrain, crops, and water availability. To manage water effectively, decisions must be made after having a thorough understanding of these systems. Here are a few popular irrigation system types:

1. Surface irrigation: This is one of the earliest and most straightforward techniques. It entails flooding or tearing up the field to create channels for water to run over the top. Despite being simple to construct, runoff and evaporation can result in water loss.

2. Drip Irrigation: Using a system of tubes and emitters, drip irrigation feeds water straight to the root zone of plants. This technique is quite effective since it uses less water and inhibits weed growth.

3. Sprinkler irrigation: Sprinklers disperse water by shooting droplets of liquid over the crops. This technique can be used on a variety of terrains and crops, but it may not be as effective due to wind and evaporation.

4. Subsurface Irrigation: Water is directly sprayed onto the root zone underneath the surface using this technique. It lessens weed growth and water loss via evaporation.

5. Furrow Irrigation: This method of irrigation uses small channels or furrows to direct water to the plants, much like surface irrigation. It's frequently utilized for row crops.

6. Localized irrigation: This technique, which provides water directly to the root zone, includes drip and subsurface irrigation. It works particularly well in places with little access to water.

7. Manual vs. Automated Irrigation: There are two types of irrigation systems: manual and automated. While automated systems employ timers, sensors, and controllers to regulate irrigation, manual systems depend on human participation to control water flow.

Basic Irrigation System Components

No of the type of irrigation system, several crucial elements must work together harmoniously to provide efficient water distribution. The following elements work together to form the foundation of an irrigation system:

1. Water Source*: This is where the water supply first started. It might be a well, river, pond, or public water source.

2. Pumping System: A pumping system is used to lift water to the necessary level if the water source is not at the desired pressure.

3. The "mainline" is a system of pipes that carries water from a source or pumping system to an irrigation region.

4. Submain: Smaller pipelines called sub mains transport water from the mainline to various fields' zones.

5. Valves: Valves control the irrigation system's water flow. They can be mechanized using controllers or operated manually.

6. Emitters or Sprinklers: Emitters are in charge of providing water to plants.

Examples of emitters include drip emitters and sprinklers. The irrigation method determines the type of emitter that is utilized.

7. Filters: Filters clear away silt and debris from the water to avoid emitters and pipelines from becoming clogged.

8. Pressure Regulators: These parts keep the system's pressure constant, ensuring that the water is distributed evenly.

9. Controllers and Timers: Automated irrigation systems employ controllers and timers to schedule and regulate irrigation cycles depending on the time of day and weather.

Determinants of Irrigation Decisions

Making wise decisions regarding irrigation requires taking into account a wide range of variables that affect the selection of irrigation systems and approaches. These variables can change depending on the region, crops, climate, and resources available. The following are some important elements:

1. Crop Type: The water needs of various crops differ. While certain crops like damp soils, others do better in well-drained soils.

2. Soil Type: Drainage and water retention are influenced by the makeup of the soil. Clay soils hold onto moisture while sandy soils swiftly drain water.

3. Climate and Weather: The frequency and quantity of irrigation are influenced by temperature, humidity, and rainfall patterns.

4. Water Availability: It's important to consider the source and accessibility of water. Effective irrigation systems are crucial in areas with limited water supplies.

5. Topography: Water distribution is impacted by the topography of the ground, including slopes and contours. While some systems may perform better on flat terrain, others excel on inclines.

6. Budget and Resources: The amount of money available for irrigation infrastructure and upkeep influences the system chosen.

7. Effort and Expertise: The operation and maintenance of some systems necessitate greater manual effort and technical knowledge.

8. Environmental Concerns: Minimizing environmental damage and ensuring sustainable water use are critical factors.

9. Crop Cycle: The crop's growth stage affects the amount of water it needs. Young plants may require irrigation more frequently than adult ones.

Conclusion

It becomes clear as we explore the principles of irrigation that it involves a complex interplay of methods, frameworks, and decision-making processes. The type of crops, the resources available, and environmental considerations are just a few of the variables that influence the selection of an irrigation system and its components. Creating effective and long-lasting irrigation techniques requires a firm grasp of these fundamental concepts. We will examine cutting-edge irrigation methods like drip irrigation and smart irrigation technology in the next chapters to deepen our understanding of water management in agriculture and other fields.

Irrigation via Drip (Chapter 2)

Drip irrigation, a game-changing development in water management, has changed the way water is given to crops and plants. This chapter will go into the complexities of drip irrigation. We'll look at its benefits and various applications, deconstruct the components and setup of a drip irrigation system, and delve into the critical areas of maintenance and troubleshooting that assure this water-saving technique's continuous efficacy.

Benefits and Applications

Drip irrigation has grown in popularity as a result of its several advantages over other irrigation methods. These benefits apply to a wide range of uses, making it a versatile choice for a variety of landscapes and crops.

Advantages:

1. Water Efficiency: Drip irrigation directs water to the root zone of plants, reducing water waste due to evaporation or runoff. When compared to surface irrigation systems, this precise application saves up to 50% of the water.

2. Reduced Weed Growth: Because water is directed to the root zones of the plants, the spaces between the plants remain relatively dry, preventing weed growth.

3. Reduced Soil Erosion: Slow and controlled water discharge reduces soil erosion caused by excessive water flow.

4. Better Crop Health: Drip irrigation keeps moisture levels steady, minimizing stress on plants and supporting healthier growth.

5. Fertilizer Efficiency: Fertilizers can be supplied by drip irrigation, providing for exact nutrient delivery to the plants.

6. Slopes: Drip irrigation works well on sloping terrains where surface runoff is an issue.

Applications:

1. agricultural: In agriculture, drip irrigation is widely utilized for row crops, vegetables, fruits, and orchards. It is especially good for water-stressed crops such as tomatoes, strawberries, and peppers.

2. Greenhouses: Drip irrigation enhances water distribution in controlled situations such as greenhouses, encouraging ideal growth conditions.

3. Landscaping: Drip irrigation is perfect for sustaining lawns, flower beds, and decorative plants, boosting landscape aesthetics.

4. Nurseries: In nurseries, drip irrigation ensures continuous and effective watering of young plants.

5. Urban Gardening: For container gardening and vertical planters, drip systems are popular among urban gardeners.

Drip Irrigation Components and Installation

A drip irrigation system is made up of many components that work together to accurately supply water to plants. Understanding these components and how they work together is critical for developing a successful system.

Components:

1. Emitters: Emitters emit controlled amounts of water. They can compensate for pressure to provide consistent water distribution across the field.

2. Tubing: Polyethylene tubing transports water from a source to plants. It is available in a variety of diameters, with smaller diameters being utilized for shorter distances.

3. Filters: Filters remove debris, silt, and particles from water, preventing emitters and pipelines from becoming clogged.

4. Pressure Regulators: These components keep the water pressure steady, ensuring that each emitter delivers the precise amount of water.

Backflow preventers protect the water supply from contamination by stopping water from flowing back into the mainline.

6. Distribution Lines: Distribution lines, often known as submarines, transport water from the mainline to various zones within the field.

7. End Caps and Plugs: These components minimize water leakage by sealing the tube ends.

Setup:

1. Site Analysis: Assess the terrain, soil type, and crop pattern of the area to find the best location for emitters and distribution lines.

2. Water Source: Connect the mainline to a water source, such as a well, a tap, or a water storage tank.

3. Filters and Pressure Regulators: To maintain clean water and uniform pressure, install filters and pressure regulators.

4. Distribution Lines: Draw distribution lines along the rows of crops or plants and secure them with stakes or fasteners.

5. Emitters: Attach emitters at proper intervals to the distribution lines. The spacing is determined by the plant's water requirements.

End Caps and Plugs: Seal the tubing's ends to prevent water from escaping.

Upkeep and Troubleshooting

Proper maintenance and fast troubleshooting are critical for a drip irrigation system's sustained efficacy. Ignoring these details might result in clogs, leaks, and uneven water distribution.

Maintenance:

1. Regular Inspections: Check emitters, tubing, and connections for damage, leaks, or blockages regularly.

2. Cleaning: To minimize clogging caused by silt and debris, clean filters and emitters regularly.

3. Flush the System: Flush the system regularly to remove any collected debris or particles.

4. Adjust Emitters: Check that all emitters are delivering the correct amount of water. Any emitters that aren't working properly should be cleaned or replaced.

Troubleshooting:

1. Inequitable Watering: If some plants get more water than others, look for blockages or faulty emitters. If necessary, adjust the water pressure.

2. Low Water Pressure: Low water pressure might be caused by a blocked filter or a faulty pressure regulator.

3. Leaks: Leaks can occur at connection points or as a result of faulty tubing. Inspect the system and repair any damaged parts.

4. Variation in Dripper Emission: Inconsistent water flow from emitters might be caused by blockage, unequal pressure, or malfunctioning emitters.

Conclusion

Drip irrigation has transformed water management in agriculture, landscaping, and a variety of other industries. Its benefits, such as enhanced crop health and water efficiency, make it a vital tool for sustainable water usage. Farmers, gardeners, and landscapers can maximize plant growth while conserving valuable water resources by understanding the components, setup, and correct maintenance of a drip irrigation system. In the following chapter, we will look at another advanced irrigation method: sprinkler systems, including their different varieties and applications.

Sprinkler Systems (Chapter 3)

Sprinkler systems are a versatile and commonly utilized irrigation method that can successfully supply water to a wide range of landscapes and crops. In this chapter, we'll dive into the world of sprinkler systems, looking at the different types of sprinklers available, the process of creating an efficient sprinkler system, and the tactics that contribute to water conservation while maximizing irrigation efficacy.

Various Sprinkler Types

Sprinklers are available in a range of styles, each adapted to unique irrigation needs, landscapes, and water distribution needs. Understanding the many types of sprinklers is critical to selecting the best system for your application.

1. Rotary Sprinklers: These sprinklers rotate in a circular or semi-circular manner, dispersing water evenly over a certain region. They are appropriate for vast lawns, athletic grounds, and agricultural fields.

2. Stationary Sprinklers: These sprinklers are installed on a spike or a base and cover a specific area without moving. They're perfect for tiny gardens, flower beds, and spot-watering.

3. Oscillating Sprinklers: These sprinklers produce a fan-shaped spray that alternately oscillates back and forth. They are appropriate for gardens and lawns and are frequently utilized for rectangular or square areas.

4. Impact Sprinklers: Impact sprinklers rotate as a result of the force of a water jet striking a set of arms. They are strong and often employed in huge areas such as farms and golf courses.

5. Pop-Up Sprinklers: When activated, pop-up sprinklers rise from the ground. Because of their aesthetic appeal and convenience, they are frequently employed in residential landscapes.

6. Gear-Driven Sprinklers: These sprinklers rotate the nozzle using gears, resulting in a uniform water pattern. They can cover greater areas and have customizable spray distances.

Creating an Effective Sprinkler System

Sprinkler system design necessitates careful planning and consideration of numerous aspects to achieve even water distribution and efficient irrigation. The following are the essential phases in developing an efficient sprinkler system:

1. Determine the Total Area to Be Irrigated: Determine the total area to be irrigated and split it into zones based on plant type, sun exposure, and water requirements.

2. Water Pressure and Flow Rate: Determine the system's available water pressure and flow rate to verify it can supply the appropriate volume of water.

3. Spacing and Overlap: Choose sprinkler kinds based on the shape and size of the region. Make sure there is enough overlap between sprinkler patterns to prevent under or over-watering.

4. Head-to-Head Coverage: This principle ensures that each sprinkler head covers the one next to it, resulting in consistent water distribution.

5. Layout and Elevation: Determine sprinkler location while taking into account elevation variations and barriers such as trees or structures.

6. Pipe Sizing: Select proper pipe sizes to reduce pressure loss as water flows through the system.

Install valves to manage water flow to different zones, and utilize a controller to automate the irrigation schedule.

8. Backflow Prevention: Install backflow preventers to keep the water supply clean.

9. Maintenance Access: Design the system with accessibility in mind to make maintenance and repair easier.

Water Conservation Techniques

While sprinkler systems provide effective water distribution, numerous water-saving methods can be implemented:

1. Zoning: To minimize over- or under-watering, divide the landscape into zones with similar water requirements.

2. Soil Analysis: Determine the ideal irrigation duration and frequency by understanding the soil's water-holding capability.

3. **Weather-depending Controllers:** Use weather sensors to change the irrigation plan depending on real-time weather conditions, avoiding watering during rain.

4. Pressure Regulation: Ensure that each sprinkler is operating at the proper pressure to avoid water misting and evaporation.

5. Mulching: Cover the soil with mulch to prevent evaporation and conserve moisture.

6. Maintenance: Inspect and maintain the system regularly to repair leaks, clogs, and misaligned sprinklers that can contribute to water waste.

7. Watering Time: Water in the early morning or late afternoon to minimize evaporation.

8. Choosing the Right Sprinkler: To avoid overspray or uneven watering, choose sprinklers that match the form and size of the area.

Conclusion

Sprinkler systems provide a diverse and effective irrigation solution for a wide range of landscapes and applications. The versatility of these systems is evident, with a wide range of sprinkler types available, from rotary to stationary. Water pressure, coverage, and layout are all important considerations when designing an efficient sprinkler system. Users can increase the performance of their sprinkler systems while avoiding water waste by implementing water conservation methods. In the following chapter, we'll look at another advanced irrigation technique called subsurface irrigation, as well as its benefits and uses in water-efficient agriculture and landscaping.

Subsurface Irrigation (Chapter 4)

Subsurface irrigation, a novel technique of delivering water directly to plant roots that functions under the soil surface, has gained popularity as a water-efficient and effective means of delivering water to plant roots. In this chapter, we will look at the physics of subsurface irrigation, the installation and management of subsurface systems, and the advantages and limits of this sophisticated irrigation approach.

The Process of Subsurface Irrigation

Subsurface irrigation works by supplying water directly to plant roots while avoiding water waste due to evaporation and surface runoff. The system is positioned under the soil's surface, allowing for effective water delivery while minimizing interaction with plant leaves.

1. ;Capillary Action: Subsurface irrigation takes use of capillary action, which is the capacity of water to travel upward via tiny gaps in soil. Water is pulled into the soil via hidden irrigation pipes and reaches the plant roots by capillary action.

2. Root Zone Saturation: The system keeps the soil wet inside the root zone, which is critical for plant development. This saturation improves plant health, decreases water stress, and lowers weed competition.

3. Reduced Evaporation: Because subsurface irrigation delivers water under the soil, it considerably minimizes water loss via evaporation, making it very effective in water conservation.

4. Lower Surface Runoff: Because water is given directly to the root zone, there is less surface runoff, which prevents soil erosion and nutrient loss.

Installation and Management of Subsurface Systems

To achieve optimum water distribution and plant development, installing a subsurface irrigation system requires careful design, accurate installation, and meticulous administration.

1. Soil study: Perform a complete soil study to determine the composition, drainage properties, and water-holding capacity of the soil. This data directs the installation procedure.

2. Spacing and Layout: Determine the distance between

irrigation lines depending on the soil's water-holding capacity and the plant's water needs.

3. Trenching or Burial: Dig trenches or bury irrigation lines at the required depth, generally inside the root zone, using specialized equipment.

4. Emitter Positioning: Install emitters at regular intervals throughout the irrigation lines to provide equal water distribution.

5. Pressure Regulation: Use pressure regulators to keep water pressure constant throughout the system, avoiding over- or under-irrigation.

6. Filtration: Include filters to avoid emitter clogging by removing silt and debris from the water.

7. Control and Monitoring: Use controllers and sensors to automate irrigation and monitor soil moisture levels.

8. Maintenance: Inspect the system regularly for blockages, leaks, and broken emitters. Adjust the system as required to account for plant growth and changing water needs.

Benefits and Drawbacks

Subsurface irrigation has several advantages, making it an appealing alternative for water-efficient agriculture and landscaping. It does, however, have several restrictions that should be addressed throughout system design and implementation.

Benefits:

1. Water Efficiency: Subsurface irrigation decreases water waste from evaporation and surface runoff, resulting in more efficient water usage.

2. Consistent Water Distribution: Subsurface systems promote consistent moisture distribution over the field by providing water straight to the root zone.

3. Reduced Weed Growth: Because water is provided under

the soil surface, weed germination and growth have less water accessible.

4. Lower Soil Erosion: The lack of surface runoff reduces soil erosion, prevents nutrient loss, and preserves soil structure.

5. Work and Energy Savings: When compared to standard irrigation techniques, subsurface systems take less work and energy to install.

Limitations:

1. Initial Cost: Subsurface system installation might be more costly owing to specific equipment and components.

2. Complexity of System Design: Proper system design requires knowledge of soil science, hydrology, and irrigation engineering.

3. Clogging Concerns: While filters may help to minimize clogging, subsurface emitters can still be clogged by small particles.

4. Root Interference: Plant roots may grow into irrigation pipes, clogging them and impacting water delivery.

5. Restricted Use: Subsurface irrigation is best suited to crops with shallow root systems and consistent water needs.

Conclusion

Subsurface irrigation is an innovative way to effective water delivery that meets plant demands while avoiding water waste. This strategy improves plant health, lowers evaporation, and eliminates surface runoff by exploiting capillary action and delivering water straight to the root zone. To achieve maximum performance, subsurface system installation and administration need careful design, correct spacing, and enough filtration. Despite its initial expenses and design challenges, subsurface irrigation is an appealing choice for sustainable agriculture and landscaping due to the advantages of water efficiency, uniform distribution, and decreased soil erosion. In the next chapter, we will delve into the world of smart irrigation systems, revealing how data-driven techniques are transforming contemporary water management practices.

Irrigation Technology (Chapter 5)

With the incorporation of technology into irrigation techniques, a new era of water management has begun: the era of smart irrigation technologies. In this chapter, we'll dig into the world of these technologies, investigating the role of the Internet of Things (IoT) in irrigation automation, the importance of weather-based irrigation controllers, and the advantages of integrating sensors to ensure optimum watering accuracy.

Internet of Things with Irrigation Automation

The Internet of Things (IoT) has transformed several sectors, including irrigation. To maximize water utilization, IoT-driven irrigation automation integrates data gathering, analysis, and real-time decision-making. This invention allows for remote control, monitoring, and modification of irrigation systems, guaranteeing optimal water distribution.

1. Remote management and Monitoring: IoT-connected devices enable customers to remotely manage and monitor their irrigation systems through smartphones, tablets, or PCs. This is particularly important for farmers and landscapers who are responsible for huge regions.

2. Data Collection: Sensors collect information about soil moisture, meteorological conditions, and plant health. This information is sent to a central system for analysis.

3. Real-time Adjustments: The system can automatically modify irrigation schedules based on real-time data, providing optimum water distribution.

4. Water Use Efficiency: IoT-powered systems minimize human mistake, resulting in less over- and under-watering and more efficient water use.

Weather-Controlled Irrigation

Weather-based irrigation controllers alter watering schedules depending on weather predictions and real-time data. These controls improve water distribution by matching it to current weather conditions.

1. Data on evaporative transpiration (ET): ET is the total water loss due to evaporation and plant transpiration. Controllers utilize ET data to calculate how much water plants need depending on weather conditions.

2. Rain and Freeze Sensors: When these sensors detect rain or freezing conditions, irrigation is temporarily halted to minimize unnecessary watering.

3. Integration with Local Weather Stations: To get up-to-date weather information and make intelligent irrigation choices, some controllers link to local weather stations or online weather services.

4. Water Savings: Weather-based controllers may help you save water by limiting irrigation during rainy periods and modifying irrigation hours dependent on the weather.

Sensor Integration for Optimal Watering

Sensors are critical for giving real-time data that allows for educated irrigation choices. They monitor soil moisture, temperature, humidity, and other environmental parameters to provide accurate and effective irrigation.

1. Soil Moisture Sensors: These sensors assess the moisture level of the soil and alert you when it is time to irrigate. They avoid over-watering by applying water only when required.

2. Rain Sensors: Rain sensors detect rainfall and halt irrigation programs briefly, preventing the system from running under wet circumstances.

3. Flow Sensors: Flow sensors measure the rate at which water flows through the irrigation system. They are capable of detecting leaks, blockages, and other anomalies, hence reducing water waste.

4. Plant Health Sensors: These sensors monitor plant health indicators such as leaf temperature and chlorophyll content, allowing watering to be adjusted to maximize plant development.

5. Wireless Communication: Because many sensors are wireless, they can interact with central control systems and make changes in real-time.

Benefits and Prospects

Smart irrigation solutions have several advantages that go beyond water conservation:

1. Water Savings: Precise irrigation results in considerable water savings, which is critical for long-term water management.

2. Time Efficiency: Automated systems save time spent manually controlling and regulating irrigation.

3. Plant Health: These technologies boost plant health by supplying ideal moisture levels, resulting in increased growth and output.

4. Environmental Impact: Water saving and decreased runoff help to preserve the ecosystem.

The future of smart irrigation promises intriguing potential as technology advances. Machine learning and AI algorithms might improve system efficiency even further by forecasting irrigation requirements based on historical data and present circumstances. Smart irrigation systems combined with smart agricultural methods may result in more complete resource management.

Conclusion

Traditional irrigation has been turned into a data-driven, automated operation thanks to smart irrigation technology. Water management procedures have been pushed to new heights thanks to the Internet of Things, weather-based controls, and sensors. These technologies improve efficiency, save water waste, and promote healthier plant development. Looking forward, the potential for ongoing innovation offers even more complex systems that connect seamlessly with larger agricultural and environmental plans. In the next chapter, we will look into precision irrigation, which combines data-driven decision-making with the skill of giving the precise quantity of water required for maximum plant development.

Precision irrigation (Chapter 6)

A major development in water management is precision irrigation, which best exemplifies the interplay between data-driven insights and efficient resource use. In this chapter, we'll delve into the nuances of precision irrigation and examine how data-driven decision-making, variable rate irrigation techniques, and a thorough understanding of crop water requirements combine to create the art of providing the precise amount of water required for healthy plant growth.

Understanding the Needs for Crop Water

Precision irrigation starts with a thorough understanding of a crop's water requirements at various growth stages. Farmers and landscapers may adjust their irrigation operations by being aware of the variables that affect water demand.

1. Crop Type: Each type of crop has different water needs. It is essential to comprehend the produced crop's particular water requirements.

2. Development Stage: During various development phases, including germination, vegetative growth, blooming, and fruiting, different crops have varying water requirements.

3. Climate and Weather: Crop water use is influenced by temperature, humidity, and sun radiation. Higher evapotranspiration rates occur when it is hot and windy outside.

4. Soil Type and Depth: Soil type and depth have an impact on water retention and drainage, affecting the frequency and length of irrigation.

5. Rooting Depth: Understanding how deeply the crop's roots encroach on the soil might assist estimate how deeply water must permeate the soil.

6. Water Holding Capacity: The soil's capacity to retain water affects the frequency of irrigation.

Techniques for Variable Rate Irrigation

To customize water delivery depending on spatial changes

within a field, precision irrigation uses variable-rate irrigation systems. By addressing particular circumstances in various locations, this method optimizes water consumption.

1. Soil Moisture Sensors: Monitoring moisture levels in real-time is made possible by placing soil moisture sensors across the field. By directing irrigation decisions, these sensors make sure that water is only applied when it is required.

2. Remote Sensing: Data on changes in plant health and water stress may be obtained from satellite images and aerial photography. Adjustments to irrigation are made using this data.

3. Geographic Information Systems (GIS): To provide irrigation recommendations for various places, GIS technology examines field topography, soil type, and other spatial data.

4. Variable Rate Irrigation Systems: Based on data inputs, these systems modify the amount of water applied, providing more to regions with greater water demands and less to places with less need.

Making decisions based on data
Precision irrigation is built on data-driven decision-making. Farmers may make decisions that maximize water distribution by combining real-time data and historical information.

1. Automated central control systems take information from sensors and weather predictions to regulate irrigation in real-time.

2. Previous Data Analysis: Predictive modeling for future irrigation demands is facilitated by the identification of trends and patterns in previous data.

3. Machine Learning : These innovations can forecast the need for irrigation based on a mix of current and past data, improving system effectiveness.

4. Smartphone Applications: With data at their fingertips, mobile apps enable farmers to remotely monitor and modify irrigation.

5. Algorithms and Models: To create irrigation plans that optimize water consumption, mathematical models consider a variety of factors.

Benefits and Prospects for the Future

Beyond only saving water, precision irrigation has other advantages:

1. Better Crop Yield: Precision irrigation encourages healthy development and better yields by precisely supplying crop water demands.

2. Resource Efficiency: Costs and environmental impact are decreased by using water, energy, and fertilizers more effectively.

3. Improved Nutrient Management: Accurate nutrient delivery to plant roots is made possible by precision irrigation, which improves nutrient uptake.

4. Minimized Environmental Impact: Soil erosion and nutrient contamination are avoided by reducing water runoff and leaching.

5. Resilience to Climate Change: By making the best use of water, precision irrigation helps plants adapt to shifting climatic circumstances.

Future possibilities are only going to get more interesting as technology advances. Precision irrigation systems might be further improved by machine learning advancements and integration with forecasting weather models, assuring the best watering schedules under a variety of scenarios.

Conclusion

Precision irrigation is an example of how knowledge, technology, and data may be used to create the optimal balance between plant demands and water distribution. The cornerstones of precision irrigation include knowledge of agricultural water needs, the use of variable-rate irrigation methods, and data-driven decision-making. This strategy improves agricultural productivity, resource efficiency, and environmental sustainability in addition to water conservation. Precision irrigation heralds a bright future in which agriculture and technology work together to provide food security and responsible resource management. We will review the various irrigation strategies in the last chapter, emphasizing the value of responsible water management in a world where water resources are limited and vital to life.

Irrigation Practices for a Sustainable Future (Chapter 7)

The adoption of sustainable irrigation practices has become critical in the face of increasing water scarcity and environmental concerns. This chapter digs into the world of sustainable irrigation, investigating tactics such as water-efficient crop selection, salinity, and soil health management, and novel solutions to water constraints. These strategies are intended to promote responsible water stewardship while also promoting productive agriculture and landscaping.

Crop Selection for Water Efficiency

Choosing crops that are compatible with existing water

resources is a critical step toward long-term irrigation. Some crops require less water by nature, making them suited for areas with restricted water supplies.

1. Native and Drought-Tolerant Plants: Choose plants that are native to the area and are naturally adapted to the climate. These plants have evolved to grow in conditions that are accessible, eliminating the requirement for heavy irrigation.

2. Xeriscaping: Xeriscaping is a landscaping practice that includes water-saving design concepts and uses drought-tolerant plants. It drastically minimizes the amount of water used in outdoor settings.

3. Crop Rotation: Strategically rotate crops to balance water needs and allow the soil to recover. Certain crop rotations can improve soil structure while reducing the requirement for heavy watering.

4. Dry Farming: Dry farming is the cultivation of crops without the use of supplementary irrigation, relying exclusively on rainfall. While difficult, it promotes water-efficient cultivation.

Salinity and Soil Health Management

Sustainable irrigation comprises more than just water saving; it also includes preserving soil health and avoiding problems like salinization that can emerge from inappropriate watering practices.

1. Salinity Management: Excessive irrigation can cause soil salinization, in which deposited salts impede plant growth. To prevent salt buildup, use procedures such as appropriate drainage and leaching.

2. Soil Testing and Amendment: Test soil quality and pH levels regularly. Improve water retention and drainage by amending the soil with organic matter.

3. Cover Crops: Plant cover crops during fallow seasons to reduce soil erosion and improve soil structure, which in turn improves water penetration.

4. Mulching: Applying organic mulch to the soil surface lowers evaporation, keeps the soil moist, and keeps weeds at bay.

Using Innovative Solutions to Combat Water Scarcity

As water scarcity becomes a global concern, novel solutions such as sustainable irrigation practices are emerging to solve this issue.

1. Rainfall Harvesting: Collecting and storing rainfall from rooftops and surfaces decreases dependency on traditional water sources.

2. Desalination: Desalination can provide an alternate water resource for irrigation in places with brackish or saline water sources.

3. Agricultural Runoff Management: Managing runoff from irrigated areas properly reduces nutrient pollution and saves water.

4. Aeroponics and Hydroponics: These soilless

cultivation methods allow for precise nutrition and water delivery to plants while reducing waste.

5. Subsurface Drip Irrigation: As previously noted, subsurface systems can reduce water loss through evaporation and runoff, especially in arid climates.

Advantages and Future Prospects

Sustainable irrigation systems provide numerous advantages that go beyond water conservation:

1. Resource Conservation: Efficient irrigation reduces the waste of water, energy, and nutrients, resulting in resource efficiency.

2. Long-Term Soil Health: Soil health and productivity are improved by practices such as cover cropping and organic matter assimilation.

3. Climate Change Resilience: Sustainable approaches enable agriculture and landscaping to adapt to shifting weather patterns and water availability.

4. Economic Savings: Farmers and landscapers benefit financially from reduced water and energy consumption.

5. **Environmental Preservation:** Good water management helps to preserve aquatic habitats and biodiversity.

The future provides bright opportunities as the relevance of sustainable practices grows. Continued research and innovation can result in better irrigation technology and tactics that improve water efficiency and environmental harmony even more.

Conclusion

Sustainable irrigation practices demonstrate a deliberate commitment to responsible water use, incorporating techniques that go beyond basic water saving. Water-efficient crop selection, soil health management, and new water-scarcity solutions all contribute to a comprehensive approach to irrigation. In a world where water supplies are limited, everyone must protect them through sustainable practices. As the trip through the complexities of irrigation systems comes to a close, the importance of balancing human demands with environmental preservation stays prominent. The path to a sustainable future necessitates a collaborative effort to assure water supply for future generations.

Irrigation Innovations (Chapter 8)

Irrigation technologies have taken center stage in the goal of sustainable agriculture and ethical resource management. This chapter explores the innovative irrigation methods that are reshaping the agricultural landscape. We'll examine the ground-breaking ideas of aeroponics and hydroponics, the potential of vertical farming and urban irrigation, and the utilization of renewable resources for irrigation. These developments have the potential to fundamentally alter how we produce crops and manage water in a world that is changing very quickly.

(Hydroponics and Aeroponics)

Two soilless growing techniques that completely rethink how plants obtain water and nutrients are aeroponics and hydroponics.

These techniques offer improved accuracy and resource efficiency by doing away with the necessity for conventional soil-based systems.

A foggy, nutrient-rich atmosphere is used to grow plants in aeroponic systems. The plant's roots hang in the air, which promotes the best possible exchange of oxygen. A thin mist of water and nutrients is directly sprayed onto the roots to maximize absorption and development.

Aeroponics uses less water than conventional soil-based systems because it provides water directly to the roots, reducing runoff and evaporation.

To provide plants with the precise nutrients they require for optimum development, nutrients are carefully regulated. Waste and runoff are reduced by this accuracy.

- **Space Optimization:** Aeroponics units may be stacked vertically, making them appropriate for urban settings with limited space.

2. **Hydroponics:** Using hydroponics, plants are grown with their roots immersed in a nutrient-rich water solution. Nutrient film technique (NFT), deep water culture (DWC), and drip systems are a few examples of hydroponic techniques.

- **Water and Nutrient Efficiency:** Hydroponics recycles water and nutrients, which results in less wastage. The exact supply of nutrients maximizes plant absorption.

- quicker Growth: Plants in hydroponic systems often develop quicker and generate better yields than those grown using conventional techniques because they have direct access to nutrients and oxygen.

Hydroponics enables year-round growth in controlled surroundings that are unaffected by the weather outside.

Vertical Agriculture and Urban Irrigation

As urbanization progresses, creative solutions for agricultural lands in cities are crucial. The difficulty of cultivating crops in constrained locations while maximizing water consumption is addressed by urban irrigation and vertical farming.

1. Vertical Farming includes growing crops at vertical levels, frequently in climate-controlled indoor settings. Vertical farms commonly employ hydroponic and aeroponic technologies.

- Space Efficiency: Using vertical space, vertical farms can grow more crops in a smaller space than conventional horizontal fields.

Vertical farming reduces water waste and runoff by carefully regulating irrigation and fertilizer supply.

- Climate Control*: Controlled circumstances make it possible to cultivate crops all year round, regardless of the weather outside.

2. Urban Irrigation: Urban settings provide particular difficulties for water management. Urban irrigation is not complete without intelligent irrigation systems, green roofs, and effective water storage.

- **"Smart Irrigation":** IoT-driven solutions allow for accurate water delivery based on real-time data, maximizing water consumption for urban landscapes.

- *Green Roofs:** Vegetation-covered green roofs mitigate the urban heat island effect and collect rainwater, promoting sustainable water usage.

- **Rainwater Harvesting:** Using collected rainwater for irrigation instead of municipal water supply helps the environment.

Irrigation with the Use of Renewable Resources

Utilizing sustainable techniques, such as renewable resources for irrigation, lessens dependency on fossil fuels and limited water supplies.

1. Solar-powered irrigation systems: These systems use solar energy to pump water. In sunny areas with limited access to electricity and water, this strategy is very beneficial.

- Energy Independence: Solar energy lessens dependency on non-renewable energy sources, reducing irrigation's environmental effect.

- ***Cost Savings:** Compared to conventional systems, solar-powered systems have reduced ongoing operating expenses once installed.

2. Wind-Powered Irrigation: Pumps for irrigation may also be run on wind energy. Water pumping is done using the power produced by wind turbines.

- **Renewable electricity Integration:** Wind power works in conjunction with solar power to produce electricity steadily even when solar radiation is low.

- **"Low Carbon Footprint"**: In comparison to conventional fossil-fuel-powered irrigation systems, wind-powered irrigation systems emit much fewer greenhouse gasses.

Benefits and Promising Future

Innovations in irrigation have a variety of advantages that support sustainable agriculture and water management, including:

1. Resource Efficiency: Water and energy usage are reduced via soilless farming techniques and systems driven by renewable energy.

2. Higher Yields: Crop yields are frequently higher when nutrients and water are delivered precisely.

3. Year-Round Cultivation: Crops may be grown all year round thanks to vertical farming and controlled settings.

4. Urban Sustainability: In highly populated places, food security and environmental issues are addressed via vertical farming and urban irrigation systems.

5. Climate Resilience: New technologies aid agriculture in adjusting to varying climatic conditions and water availability.

The future presents even more interesting possibilities as technology continues to advance. Automation, data analysis, and the integration of renewable energy sources might transform irrigation techniques, furthering efficiency and sustainability.

Conclusion

Irrigation technology advancements are altering the agricultural environment, fostering resource efficiency, and confronting issues with water shortage head-on. The ability to grow crops sustainably in a world that is changing quickly is demonstrated by the use of systems driven by renewable energy, aeroponics, hydroponics, vertical farming, and hydroponics. As the chapter on innovations draws to a close, the larger story of responsible water stewardship and sustainable agriculture stays in the foreground. The voyage of irrigation technology evolution highlights the significance of adjusting to and accepting new practices that balance environmental protection with human demands.

Case Studies (Chapter 9)

Real-world examples of sophisticated irrigation systems successfully implemented give significant insights into the practical effect of current irrigation technologies. In this chapter, we will look at three different case studies: high-yield farms that benefit from contemporary irrigation, urban landscapes that thrive with effective watering procedures, and desert land rehabilitation using smart irrigation tactics. These case studies demonstrate the transformational effect of novel irrigation systems in a variety of settings.

Case Study 1: Modern Irrigation on High-Yield Farms

Modern irrigation systems have played a critical role in the goal of sustainable and high-yield agriculture. This case study focuses on a major agricultural operation that used precision irrigation, subsurface irrigation, and data-driven decision-making to maximize crop output while saving water resources.

Scenario: A 500-acre farm growing a wide range of crops.

1. Precision Irrigation: Precision irrigation was established on the farm by combining soil moisture sensors and weather-based controls. The sensors' data was transmitted into a central control system, which altered watering schedules depending on real-time weather conditions and soil moisture levels.

2. Subsurface Irrigation: In places where water shortage was an issue, a subsurface irrigation system was developed. This technique reduced water waste caused by evaporation and surface runoff.

3. Data-Driven Decision Making*: Historical crop water needs, growth phases, and yield data were examined. Machine learning algorithms projected watering requirements and guided automatic system modifications.

Results:

1. Increased Yield: Precision irrigation and data-driven techniques resulted in optimum irrigation and increased crop yields.

2. Water Savings: Thanks to accurate irrigation schedules and subsurface systems, water use was decreased by 20%.

3. Resource Efficiency: Nutrient supply is coordinated with water distribution, boosting plant nutrient absorption.

Case Study 2: *Efficient Watering Promotes Urban Landscape Growth*

Sustainable water management is essential for dynamic urban settings. This case study looks at a city's attempts to revitalize its parks and public areas via the use of smart irrigation systems.

Scenario: A city that is experiencing water shortages and wishes to increase the efficiency of its irrigation systems.

1. Smart Irrigation Systems:
The city modernized its irrigation systems with Internet of Things-powered controls and soil moisture sensors. These devices gathered data on soil moisture, temperature, and weather predictions in real-time.

2. Weather-depending Adjustment:
Irrigation schedules were altered depending on actual

and expected weather conditions, eliminating needless watering during rain or cold spells.

3. Drip Irrigation Conversion: In flower beds and landscaping, traditional sprinkler systems were replaced with drip irrigation. This reduced water waste from evaporation and overspray.

Results:

1. Water Conservation: The city cut its irrigation water consumption by 30%, easing the burden on local water supplies.

2. Landscape Health: Because of appropriate irrigation, plants flourished, resulting in healthier and more aesthetically pleasing urban landscapes.

3. Cost Savings: Lower water use resulted in lower city utility expenses.

Case Study 3: *Arid Land Rehabilitation Using Smart Irrigation*

Smart irrigation systems may play a critical role in land rehabilitation in dry climatic zones. This case study investigates the use of underground irrigation and weather-based controls to turn a degraded area into a productive ecosystem.

Scenario: A semi-arid village suffering desertification and soil deterioration.

1. Subsurface Irrigation: Water was delivered directly to the root zone of natural vegetation and strategically planted crops using subsurface irrigation lines.

2. meteorological-Based Controllers: Irrigation schedules were generated automatically by combining local meteorological data and evapotranspiration rates.

3. Native Plant Restoration: The community launched a native plant restoration initiative, using the subsurface irrigation infrastructure to care for newly planted plants.

Results:

1. Land Restoration: Degraded land was turned into a lush environment with better soil structure and plant cover.

2. Desertification Mitigation: By generating favorable circumstances for plant development, smart irrigation technologies slowed the progress of desertification.

3. Community Resilience: The restored area offered animal habitat and increased local biodiversity.

Conclusion

The case studies in this chapter demonstrate the many uses and advantages of sophisticated irrigation systems. These real-world examples highlight the revolutionary potential of current irrigation systems, ranging from high-yield farms that optimize water usage to urban landscapes that thrive with efficient watering and the rehabilitation of desert regions with smart irrigation. These case studies serve as templates for implementing new methods that balance agricultural and environmental demands, emphasizing the significance of sustainable water management in constructing a resilient and productive future.

Future Irrigation Trends (Chapter 10)

The development of irrigation continues to unfold as the globe moves toward a future marked by population expansion, climate change, and resource constraints. This chapter dives into future developments that have the potential to change irrigation techniques. We will look at how predictive analytics may be used in irrigation management, the role of nanotechnology in water distribution, and the importance of collaborative approaches to water sustainability. These developments hint at the revolutionary potential of irrigation methods in the next years.

Irrigation Management Predictive Analytics

Data analytics, machine learning, and irrigation management together can transform water distribution methods. Predictive analytics use historical and real-time data to forecast irrigation requirements, resulting in more efficient and informed decision-making.

1. Data Integration: Predictive analytics collect information from a variety of sources, such as weather predictions, soil moisture sensors, and crop growth trends.

2. Machine Learning Algorithms: Machine learning algorithms analyze this data to detect patterns, correlations, and trends, allowing future irrigation needs to be predicted.

3. Automated Adjustments: Predictive models may alter irrigation schedules automatically depending on predicted weather conditions and crop water demands.

Benefits:

1. Increased Efficiency: Predictive analytics allow for exact irrigation scheduling, which reduces water waste and over-irrigation.

2. Resource Optimization: Aligning water, energy, and

fertilizer consumption with crop needs optimizes resource utilization.

3. Climatic Change Resilience:
The flexibility of prediction models aids agriculture in dealing with changing climatic circumstances.

Water Distribution and Nanotechnology

Nanotechnology, or the manipulation of matter at the nanoscale, presents the fascinating potential for altering irrigation water delivery.

1. Water Filtration Nanomaterials: Nano-sized particles may be employed in filters to remove toxins from irrigation water, hence increasing water quality.

2. Smart Irrigation with Nanosensors: Nanoscale sensors provide extraordinary accuracy in detecting soil moisture, nutrient levels, and other characteristics.

3. Irrigation Pipe Nanocoatings: Nanocoatings may reduce friction in irrigation pipes, lowering energy needs for water delivery.

Benefits:

1. Precision Water Management: Nanosensors provide real-time information on soil conditions, allowing for precise irrigation changes.

2. Improved Water Quality: Nanotechnology enhances water filtration and purification, protecting crops from pollution.

3. Lower Environmental Impact: Nanocoatings and filters increase irrigation system efficiency, minimizing water and energy waste.

Water Sustainability Through Collaborative Approaches

To address water shortage and sustainability, stakeholders such as governments, farmers, academics, and communities must work together.

1. Data Sharing and Management: Collaborative systems enable farmers and institutions to share data, allowing for more informed decision-making.

2. Integrated Water Management: Collaboration between agriculture, industry, and the urban sector aids in optimizing water utilization across several demands.

3. Community Engagement: Involving communities in water

management encourages responsible usage and promotes long-term habits.

Benefits:

1. Holistic Resource Management: Collaborative efforts guarantee effective water allocation, which benefits both agriculture and society.

2. Local Solutions: Communities may collaborate to develop solutions that are suited to their unique water availability and demands.

3. Adaptation and Resilience: Collaborative methods improve the ability to adjust to water shortage concerns and changing circumstances.

Conclusion

Irrigation's future is formed by a dynamic interplay of technology developments, new methods, and collaborative efforts. The combination of predictive analytics, nanotechnology, and collaborative water management systems has enormous promise for revolutionizing irrigation operations. As the world grapples with serious issues such as population increase, climate change, and water shortages, these future patterns provide a ray of hope—a road map for sustainable and resilient water management. Irrigation method development is an important aspect of humanity's quest to balance its demands with the environment, encouraging a future in which water resources are nurtured, conserved, and shared for the benefit of everyone.

Conclusion

One overarching theme emerges from the voyage across the complicated universe of irrigation systems, from historic traditions to cutting-edge innovations: the requirement to balance our relationship with water resources. The conclusion of this investigation invites us to acknowledge the transforming potential of contemporary irrigation and our common duty in crafting a greener, more sustainable future.

Moving Towards Modern Irrigation for a Greener Future

Irrigation's development reflects humanity's growth and progress, demonstrating our capacity to leverage knowledge and technology for the benefit of both society and the environment. Modern irrigation methods, ranging from precision irrigation to smart systems, solve water shortages, increase agricultural productivity, and contribute to responsible water management.

This tour emphasizes the need of adopting contemporary irrigation methods as a cornerstone of sustainable agriculture and landscaping. Water, a scarce and priceless resource, requires careful management and inventive solutions to assure its availability for future generations. We can enhance agricultural yield, preserve water, and reduce environmental consequences by using contemporary irrigation systems.

Your Role in Flow Mastery

As we come to the close of our investigation, it is important to remember that the road toward responsible irrigation is a joint effort including people, communities, corporations, and governments. Each of us has a critical role to play in controlling the flow of water for a greener future:

1. Farmers and Landscapers: Using contemporary irrigation technologies improves yields while also conserving water,

lowering expenses, and promoting sustainable land management.

2. Researchers and Innovators: Your quest for knowledge and invention promotes irrigation method development, ushering in a new age of resource-efficient practices.

3. Governments and Policymakers: You play an important role in adopting policies that encourage sustainable water management, incentivize the use of contemporary technology, and support joint efforts.

4. Communities: By practicing water-wise habits and supporting local efforts, you help to promote responsible water use and environmental resilience.

5. Consumers: Consuming water-intensive items and foods mindfully promotes knowledge about water footprints and supports sustainable choices.

The Following Chapters

The tale of contemporary irrigation is far from over; it is a never-ending one of adaptation, invention, and harmonization. As the globe grapples with the issues of a changing climate, expanding population, and limited resources, irrigation systems will continue to evolve. The future is unexplored land, blending accuracy and intuition, data and knowledge, and innovation with tradition.

We have a unique chance to design a more sustainable world

as we stand at the crossroads of previous practices and future trends. Individual and community collective efforts, the devotion of academics, and the vision of politicians will write the chapters that follow. Every innovative irrigation system built, every drop of water conserved, and every collaborative effort made adds to the story of a greener, more resilient future.

In conclusion

Irrigation is more than just watering crops; it is a metaphor for our interaction with the environment and the conservation of resources that support life. The trip through these chapters emphasizes how water is inextricably linked to the flow of development, knowledge, and responsibility. As individuals and as a community, let us hear this call to action and embrace contemporary irrigation expertise for a future where water flows responsibly, landscapes thrive vibrantly, and the fabric of life grows harmoniously.

Key Terms and Definitions Glossary

The study of this crucial technique is supported by a wealth of specific vocabulary and concepts in the field of irrigation. To help readers understand the wide range of irrigation principles and procedures, this glossary offers a succinct yet comprehensive collection of key terminology and their definitions.

1. Irrigation: The deliberate application of water to soil, vegetation, or crops to promote growth and productivity.

2. Evapotranspiration: A process that involves both plant transpiration and water evaporation from soil and water surfaces.

3. Water Scarcity: A situation in which a region's need for water outweighs its supply.

4. Drip Irrigation: This approach uses a network of pipes, tubes, and emitters to deliver water directly to plant roots while minimizing waste.

5. A sprinkler system is an irrigation technique that disperses water through a network of pipes and nozzles to simulate rainfall.

6. Subsurface irrigation: This irrigation method applies water below the soil's surface to minimize evaporation and runoff.

7. Precision Irrigation: A focused method of water distribution that makes use of data and technology to maximize water use.

8. "Salinity" is the number of salts in soil or water, which can have an impact on a plant's health and growth.

9. Aeroponics: An approach to soilless gardening in which plant roots are suspended in the air and water and nutrients are provided as a fine mist.

10. Hydroponics: A technique for growing plants without soil in nutrient-rich water solutions.

11. Vertical farming: Growing crops in layers that are piled vertically, frequently indoors, to make the most of available space.

12. The Internet of Things (IoT) is a network of connected objects that can exchange information and communicate with one another online.

13. Nanotechnology: The tinkering with matter at the nanoscale to produce new products and methods.

14. "Desalination" is the process of purifying seawater or brackish water to remove salt and other pollutants and make it fit for drinking or irrigation.

15. Rainwater Harvesting: Gathering and preserving rainwater for use in irrigation and other applications.

16. "Collaborative Water Management" refers to a group effort including various stakeholders to sustainably manage water resources.

17. Predictive Analytics: The application of data analysis and modeling methods to forecast upcoming occurrences or trends.

18. Nanosensors: teeny, tiny sensors that can gather and communicate data on a variety of parameters, such as soil moisture and nutrient levels.

19. "Climate Resilience" refers to a system's or community's capacity for adaptation and success in the face of changing climatic conditions.

20. "Resource Efficiency" is the best use of resources like water, energy, and nutrients to reduce waste and increase output.

21. Soil Moisture Sensors: Instruments that gauge soil moisture levels and inform irrigation decisions.

22. The "Nutrient Film Technique" (NFT) is a hydroponic technique that involves circulating a thin layer of nutrient-rich water over plant roots.

23. Deep Water Culture (DWC) is a hydroponic technique that suspends plant roots in a nutritional solution.

24. Green Roof: A roof covered in flora that manages stormwater runoff, conserves water, and acts as insulation.

25. Machine learning is a branch of artificial intelligence that enables computers to learn from experience and advance without explicit coding.

Renewable resources, such as solar and wind energy, are resources that replenish themselves spontaneously and have no impact on the environment.

27. Deforestation, drought, and bad farming techniques are some of the elements that cause productive land to become desert.

28. "Xeriscaping" is the practice of landscaping and gardening that uses water-wise plant selections and design ideas to minimize the need for additional water.

29. Urban Heat Island Effect: A condition whereby urban areas heat up more quickly than nearby rural regions as a result of infrastructure and human activity.

Making well-informed decisions in light of data analysis and interpretation is known as data-driven decision-making (30).

Conclusion

Similar to how water flows, irrigation has a complicated and diverse vocabulary. This glossary's terms each stand for a stitch in the tapestry of ethical water use and environmentally sound farming methods. May this glossary prove to be an invaluable resource as you set out on your quest to master irrigation's art and science, helping you to understand the nuances of contemporary irrigation methods and preparing you for the opportunities and difficulties that lie ahead.